SOTHIS PRESS

Sothis Press aims to completely reimagine the public domain book with an emphasis on universal access to captivatingly designed print editions. We believe that technology will enable every single book ever published to be perpetually accessible in beautiful print form. Sothis is beauty in ubiquity.

www.sothis.press
1 Waterhouse Square, London, ECIN 2ST

The Sand Dunes of Indiana

The Story of an American Wonderland
Told by Camera and Pen

Eli Stillman Bailey

SOTHIS PRESS

ISBN 978-1-0368-0771-9

The Sand Dunes of Indiana
is in the public domain.

This edition is published by
Sothis Press © 2023. The company
is registered in England and Wales.
The Company Number is 14900389.

THE SAND DUNES
OF INDIANA

The story of an American wonderland told by camera and pen

by

E. STILLMAN BAILEY

PROFUSELY ILLUSTRATED

CHICAGO
A. C. McCLURG & Co.
1917

Published December, 1917

To

A. W. P.

A LAND OF DREAMS

'The' Dune Country of Indiana represents the work of one hundred times one thousand years, by such artists as glaciers, water, wind, and sun, until you find there a park, perfect, beautiful; a fairy land; a land of dreams; a land of remoteness; a land of solitude; a land of long beaches; a land on whose frail shores strong waves beat at times with thunderous roar; a land so fair and fine no city park could be made equal it by the expenditure of countless millions."

—THEODORE JESSUP.

PREFACE

THE sand dunes of Indiana are one of the wonder places of America. They are well known and yet unknown. For many, many years the great traveling public, when going either east or west, to or from Michigan City, Indiana, have glanced casually from the car windows, as they passed the dune district, at the enormous sand piles. They saw the dunes, it is true; but to see them in this manner is simply to place them geographically. It is to further a greater intimacy with the dunes that these pages have been written. I want the thousands of people residing near them to know something about them—what they are, and how they were formed.

At the present time there is a rather limited number of books descriptive of sand dunes to be found in the large circulating libraries. I have consulted some of them, and credit is cheerfully given those from which I have made quotations.

The illustrations are reproductions from photographs selected from a large number of sand-dune views of my own collection. For the most part they illustrate the various forms of sand dunes, and in a measure I have tried to make them tell the story of the dunes.

E. S. B.

The crumbling trunk and tap roots
of a once stately white pine.

CONTENTS

THE LURE OF THE DUNES

In Dune Land there are miles and miles of scenes like this.

The Sand Dunes of Indiana

CHAPTER I

THE LURE OF THE DUNES

A DUNE is a pile or ridge of incoherent sands, orderly or fantastically fashioned by prevailing winds. Sand dunes are common the world over; they are inhospitable spaces on the earth.

The lure of the dunes is kin to the lure of the desert and the desire for great waste spaces. They have, however, a uniqueness of their own which admits of no comparison. They breathe mystery and romance, and appeal to the imagination. They have a temper and charm which you come to know and feel. At times they are in the area of fierce storms, with desert winds and blinding sands, and at other times when calmness obtains, one can, paradoxically speaking, listen to the voices of their silence. A hundred theories may explain their appeal, but, after all, the spell they cast becomes your own to unriddle, and you answer the riddle in your own way. The dunes are alluring and fascinating, and with the varying sea-

A typical view in Dune Land. Photograph taken from the crest of a wandering dune near the Mineral Springs road.

The skyline from an elevation near Dune Park, showing the dunes of that region in the distance. The view is very beautiful from this spot, the contrast of the white sand hills in the background with the dark green of the vegetation near-by making a most charming effect.

sons their changing forms and colors weave for you a wealth of fact and a wreath of fancy. Their secrets are many, and their wild beauty is a pleasure to the eye and a joy to the memory.

Perhaps it is the out-of-doors, the unconstraint, the

Where the sky, sea and sands meet.

farewell to bricks and traffic, that fascinate you at the first visit, for fascinated you will be. Maybe it's the beach, where the sky, the water, and the sand make a long blue line, with miles of freedom to enjoy. Perhaps it is the place where work and care are tossed upon the crest of the outgoing wave, and time is immersed in forgetfulness. Maybe it is the exchange of the plain or prairie for the hills, for before you is the skyline of

A typical picture of one of hundreds of beautiful paths through the woods of the dunes. Photograph taken near Michigan City.

A familiar scene along the beach, showing fishermen's nets and the warehouse in the foreground. The fishing season is short and the work is hard.

the heights, a hundred or more feet above you, of changing sands, changing forms, changing views, and exchange of colors with a decoy that tempts you to tread an unbroken path in the mobile sand and climb

The very bleakness fascinates.

to the top of the unstable crests. Whatever it may be, visits to the sand dunes are unconventional and they beget and foster a wanderlust.

The dunes are mutable, the mobile forms of yesterday pass into oblivion while tomorrow's are in the making. The dunes are matchless, their class is all their own; they are restless, changing, billowy sands.

Freedom pervades the whole atmosphere of the dunes. The voices there seem to say, "You do not trespass, you are welcome, you are no bother, you must take us as you find us today, for this is Utopia." If you care to sit and dream, the foundations for your air castles are

The old and the new. In the foreground are yearling seedlings and in the background a century-old dead forest.

here. If you enjoy the sudden change from your home to this wilderness, you will return again and again, and in time be recognized as a regular dune commuter and an all-the-year-round visitor.

To the query as to the best time to see the dunes, I reply, the best time is at your own convenience. To some, the dunes are uncomely—yellow, bothersome,

Nature strives with Nature in Dune Land. The cottonwoods are struggling, even changing their type in the effort, to exist. To the right the grape vine is trying to hold the hill of sand together. In the foreground the blow-sand is rapidly coming in and in a few years the trees and vines will be buried and the scene changed entirely.

Winter in the Dune Country. The dune shown in the picture is an old one of the fixed or dead type; many forest fires have swept over it. Photograph taken about half a mile south of Sheridan Beach by Dr. W. H. Wilson.

changing sands, and nothing else. The gift of imagination was not bestowed on these people. They are to be pitied, for they miss much of life's finer pleasures.

Long Lake in autumn. The colorings were wonderful.

To others, the dunes are fascinating at any and all times. Perhaps, as the "old commuter" said to me recently, the dunes are at their best in the spring; but the same enthusiast must have a short memory, for last fall, while his arms were laden with American holly and bunches of bitter-sweet for his home holiday decoration, he told me that the fall of the year is the best possible time for these excursions. So you see any time is right if *you* choose it.

If you are feeling fit, and are warmly clad, you will welcome a trip to the dunes even during the winter's snows and gales, and find a rich return for your venturous jaunt.

If you are the Indian you think yourself, you will on a summer day take a fifteen-mile hike on the beach, hatless, and unconventionally plunge at your own will into the lake for a refreshing swim, and later you may seek a resting spot to watch the sunset; or, after a night in a shack or under a bent tree for shelter, with a blanket for a cover and the warm sinking sand for a bed, you may see the silhouette of the sentinel pines against the morning sky, conscious of the beginning of another day of moods among the dunes.

If you are hopelessly practical, a one-day visit will suffice for you, but if you are thrilled by something new or gifted with imagination, or if you are curious and temperamental, the call of the sand dunes will be resistless.

Some day you may happen at the dunes when the silence will be the charm. The wind has ceased, the little lisp of the water as it caresses the sand will be the only sound. The surface of the lake will be molten. It is a calm that may cast the spell of littleness of self, and suggest the largeness of nature's forces. The calm at the dunes is to be felt; it cannot be spoken. The sky, the water, and the land meet and proclaim the peace of heaven and earth and sea.

The next day it may be the fury of the wind that will cast its spell upon you. It is well to be prepared

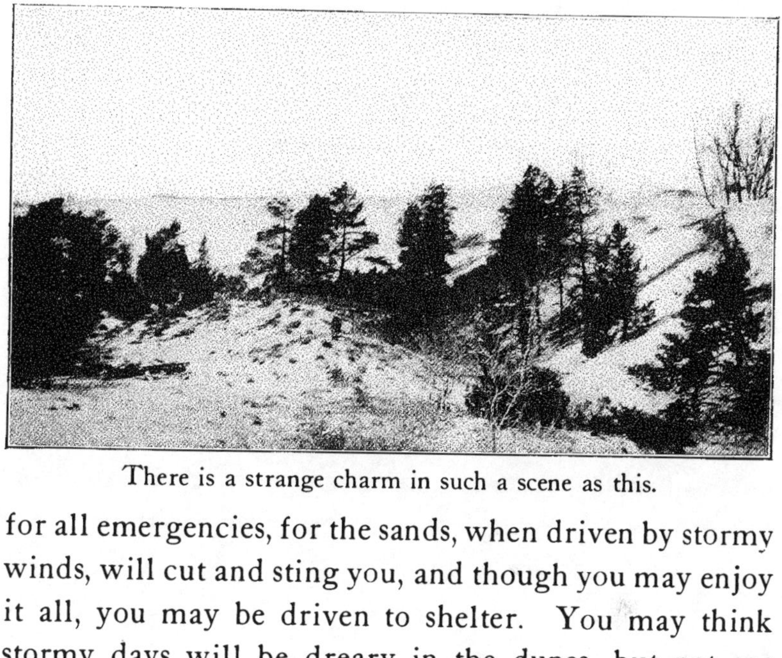

There is a strange charm in such a scene as this.

for all emergencies, for the sands, when driven by stormy winds, will cut and sting you, and though you may enjoy it all, you may be driven to shelter. You may think stormy days will be dreary in the dunes, but not so; both storm and calm are but echoes of your own self; you acquiesce when all is still, and you thrill when all is moving.

It is quite unnecessary for you, in order to enjoy the dunes, to be a geologist or a botanist or a zoölogist, or to know about the formations or historic relationship of glaciers, ice-fields, sedimentary rock, granites, beach lines, or ocean beds. It is all the more interesting if you happen to possess this fund of information, but if you can see and feel the real dune pictures, with their high lights and contrasting colors, the shadows and shades of the purples, blues, grays, yellows, and greens, whether of the lake, the sands, the clouds, the mists, or the receding paths into tempting woodlands, or views where land and water meet, and if the vistas spread before you charm your eye and rest your brain, then the dunes are for you.

Every branching tree or wild vine, every print of foot of tern or turtle, every lee of rotting wrecks or upturned roots, every little ripple on the beach or on the sanded crests, every beating wave when the storm is wildest, or whisper of wind in the forest depth, or note of bird or tread of hoof, every sandy trail to flower or fern, every echo from surf or camp will cast its mystic spell and make a memory web. If you see these views or hear these things, or feel it all, you may be certain that the dunes are for you.

At your feet even the rounded smooth stone may suggest a story. Maybe it is a wanderer from some far-distant home, where other sands have fashioned its form and brightened its face. If you pause and gaze at the sentinel pines that stand, even when dead, pointing their tops above the advancing sand; if you see the polish

The flora of the dunes is marvelous. Here the wild cherry trees, tall pines, low conifers, cedars and shrubs and pine undergrowths are in glorious profusion. The dune binding grasses are also present in the foreground. The picture is typical of the flora of the windward (northern) side of the dunes.

View taken near Dune Park from the rim of a big blowout on the beach side of the dunes. The time was autumn and the splendor of the rich colorings of the forest was magnificent.

on the fallen trunks and roots of trees, made by nature's sand blast of prevailing winds and diamond sand dust, then, too, the dunes are for you and are yours to enjoy. If your camera takes what your eye sees, the pleasures

In the grasp of winter.

will double as you review the pictures from time to time in your leisure hours.

The call of the sand dunes is the call of your measure of fancy and of caprice. There is always the riddle of varying attractions, and each journey to the dunes has a reward equal to that of a new discovery. If your first visit be in the fall of the year, the views of the autumn will outrival those of the spring. Then the dunes are a blaze of colors which are heightened by the white of the sands and mellowed by the contrasts of evergreens and ripening vines. The rushes and grasses may have withered, but the autumn trees will command your attention, for so red are the maples, so brown the oaks, so yellow the aspen, and so green and feathery the tamarack, that one draws near to see it all. If the frost should have completed the ripening of the woodbine and ivy on the stumps of trees or over the rotting fences, the near view will be the more charming one. The fall of the year, too, offers a chance to walk and scuff through the fallen leaves or to pick the wild grapes, whose color is now deep purple and whose bloom is now complete.

The Indian Summer days are rare days in the dunes, with their purple haze, the curling smoke of camp fires, the calls of the wild fowl returning to the lake and to the marshes, the thrill in the atmosphere, and then the stillness over it all!

The dunes are just "different"; their language is beginning to be a different tongue, their call is irresistible. You may have come impelled by curiosity, but the wild beauty invites you to remain. And why should the despoiler come? It is time now to change the waste into a park, and the desert into a playground for millions forever to enjoy.

THE FORMATION OF THE DUNES

A sea of sand.

CHAPTER II

THE FORMATION OF THE DUNES

DUNE comes from the term *dun,* which is of Celtic origin; *dun* meaning "hill." By usage the word became *dune.* Theoretically, any ripples of sand caused by a wind to change their position and pile up into a hill are dunes. It is a dune just the same if its size be such that a kick will remove it, or if it be a mountain of sand challenging a tempest to displace it. For the purpose of this book, dune means a sand hill.

Sand formations in general are named: *Aeolian,* where the sand formations are due to the agency of the wind; *Neptunian,* where the formations and sediments have been built up by the water. The transporting power of water is vastly greater than that of the wind.

Dunes are formed in lines transverse to wind in unobstructed places, and are formed in lines parallel with wind in places where some kind of obstruction is in the path of the wind.[1]

Dunes, for purpose of description, are technically called gray or stationary, and white or shifting. The

[1] Pehr Olsson-Seffer, *Genesis and Development of Sand Formations on Marine Coasts,* Augustana Book Concern, library publication No. 7. Rock Island, Ills.

terms live, active, wandering, new, traveling, are used to refer to shifting sands.

The names dead, fixed, old, stationary are used to designate dunes that no longer change. Their sandy sides are overgrown with binding foliage, and from distances they no longer show a glistening white. They are known along all seacoast lines as gray. Some call the shifting sands, slips; and the fixed dunes, knobs. The active or wandering sand heaps and ridges are those which are either forming or disappearing. The change may take place very slowly or it may be remarkably rapid. Every typical shifting dune has its convex side facing the direction of the prevailing winds. If the winds are from the west, the sand under its influence wanders in an easterly direction.

It is paradoxical that the active or live dune shows us a picture of desolation, while the dune that is alive with vegetation — trees, grasses, even flowers — is a dead one; for when a dune becomes stationary it dies, it is no longer active, the sand is being held in check, it cannot drift. The living, grass-covered and vine-clad appearance of a dune is its own tombstone. All of the extreme southern end of Lake Michigan, along praetically the whole coast between the city of Gary and Michigan City, a distance of twenty miles or more, is marked at the present time with active or traveling dunes. The waves in the time of storms have thrown

A sand dune near the shore of Lake Michigan west of Michigan City. Formerly it was of great beauty, but the winter gales have almost stripped it of its once picturesque appearance from the water side. It is very bold and very steep.

immense quantities of sand on the shore, and the wind has carried it and piled it in dunes and ridges varying in height from a foot or two to a hundred feet or more.

A dead dune is practically a heap of sand held in place by the vegetation by which it is covered.

At the time of the formation of the heap it was, of course, an active dune, but for some reason the active heap ceased to obey the winds' will, vegetation appeared, as vegetation always will if given a chance, and the dune became what is known as "fixed," or "dead."

The ridges are also dunes gone to rest, they take on more regular forms and between them there are depressions—ravines, swales, swamps, and fresh-water marshes, and ponds. The undulations of the ridges are clothed with grasses, vines, shrubs, and trees, and apparently they have taken on permanent form; but if a disturbing cause, such as a tornado, the grading for a new railroad line, a clearing for a foundation for a house or barn, or a mechanical cause of any kind, opens up a path for the wind, and the verdure is swept away, the dune becomes active at once, and its influence is not alone on that single spot; for the drifting sands are the forbears of a new dune, and the wind carries the sand wherever it listeth.

Sand dunes are held together and made to assume

A type of protected sand dune. Note the binding grasses at the base and crests. Some destruction is shown, but the vegetation is holding the dune together.

The doomed sentinel. In January, when this picture was taken, this lone tree stood erect so proudly and sturdily on the storm-swept ridge that I called it "The Sentinel." Alas, it was doomed, however, for when I visited the place later the spring winds had accomplished its downfall and changed the entire scene.

somewhat characteristic forms by their gravity, and partly by the organic material that has been deposited from various outside sources. There is also possibly a slight cohesion of the particles of lime, clay, and organic forms mixed with the sand, and by water governed by capillary attraction or by evaporation from the lower

Trio of turreted dunes.

strata, and by the retention of rain water. The moisture of dunes is characteristic, and plays a very great part in the support of vegetation and the maintenance of their form.

Streams thread their way between these sand ridges in a leisurely manner, for their drainage into some open sea is uncertain; their outlets are sometimes obliterated by changing sand bars.

A certain form of the traveling dune is called a blowout. Whenever the protective covering of vegetation is once broken, sand blowing proceeds rapidly, resulting in a saucer or bowl-shaped excavation known as a blowout. Sometimes the excavation is called a sand amphitheater or a horseshoe dune. Each is a fairly descriptive term.

The size of a blowout is consistent with several facts. The sides usually are fashioned into a gentle slope, while the crests are stopped by some growth that restricts the blowing, usually sand-binding grasses. The depth may be as deep as the sand itself, and the entire area may cover tracts several acres in extent, or be a very simple saucer-like affair and of small size, though characteristic in shape.

The sand removed by blowouts is piled up on the leeward side in an orderly fan-shaped heap, and this in turn is blown on by winds and takes the characteristic of a traveling dune. Blowouts in turn may again be filled, or partially so, by sand, being thus transformed into level tracts. This surface is then termed, blowsand, and when it becomes large the individual blowouts lose their identity and the whole tract becomes a vast undulating surface of shifting sand. The limit of size is reached when the blowout becomes so large that it no longer offers much resistance to the wind, or so deep that the wind does not have sufficient force to

A dune showing the steep side lines of a large amphitheater or blowout. The walls here are deeply grooved. Note the long roots exposed by the winds.

carry the sand from the bottom up over the sides, or where moist layers of sand are reached which cannot be blown. Sometimes real amphitheater forms are fashioned. The shape is always dependent upon some obstructive form of vegetation.

Professor H. A. Gleason says:

> In young blowouts where the excavation is being carried on most rapidly, vegetation is very sparse and the few species able to grow in such conditions constitute the typical blow-sand association. They are mostly plants with a short period of development which may mature before the shifting of the sand has undermined their root system, and they frequently possess methods of seed distribution by which they are enabled to colonize rapidly on barren areas of sand.
>
> Cassia prefers sand that is loose from blowing or that has been otherwise disturbed. Wagon tracks across the sand are quickly occupied by it, and are marked by long parallel lines of the plants which are very conspicuous in the blowing season.[1]

Following blowouts, vegetation is always scattered, and the blowing of sand during the fall and winter is very apt to prevent perennials taking any part in sand binding.

Cottonwood seedling trees grow rapidly in these sands. Their location is accidental. They spring up singly or in thickets, according to the planting. These broad-leaved trees adapt themselves to the barren zones as readily as do the grasses. Cottonwood trees are about the best sand binders, owing to their extensive root sys-

---

[1] *Bulletin of the Illinois State Laboratory of Natural History*, 1909, p. 149.

tems. They demand moisture, and the rootlets will travel far to get it. Should the trunk and limbs be overwhelmed with sand, the sand-covered limbs will quickly change to roots and the trunk will push

A thicket of thriving cottonwoods.

up rapidly; thus an additional anchorage is added. Equally important are these trees in the prevention of the wind destruction of the dunes. The cottonwoods maintain a vigorous growth and tenacity in these seemingly impossible places.

The sand removed from each blowout is distributed over a larger area than that from which it was taken, so that the tendency of wind action is not to produce large drifts of sand or to fashion steep sides or crested tops, but to act toward a larger level surface of sand. It must be remembered always that the wind is the instrument of excavation, and a peculiarity of wind-blown sand is that it does not rise like dust, but flows forward, suggestive of water.

It takes a little time to become intimate with the amphitheaters of sand, or blowouts. Usually they present a front of several hundred feet. The side presented to the wind was once hard-packed sand and had upon it little or no vegetation. An opening is made by some cause, the dried sands are hurled backward over a long distance—a fourth of a mile is not uncommon—the immediate front being scooped out, leaving abrupt side walls, with angles sometimes equal to seventy-five degrees. The scooping-out process keeps on until the floor of the dune offers no further resistance; the dune has been reduced to a gentle slope of ten to twelve degrees, its sand is away to the leeward, spread out at an angle of thirty to sixty degrees. If you try to walk in it you sink to your knees.

Such an amphitheater can be easily studied—there is one at Carr's Beach (Miller), extending back to the bridge. The scooped-out part in the center at the present time is being anchored against future destruction by a growing forest of cottonwood seedlings. The board

The bridge which spans the Calumet River at Miller. The board walk leads to Carr's Beach, a quarter-mile north. The dune on the right is of the blowout type.

walk is nearly in the center of this blowout. Its length is the length of the wood walk north and south, and its width are the two high crests, one to the east, the other to the west. It is a typical dune of its class.

The so-called "blow-sands" are the mobile sands that move with the breath of the wind or scurry with the strength of the gale. They are the fine dry sands, always on the surface of things, and are as fickle as the wind which carries them.

Some dunes are called complex dunes, because, although they travel mainly in one direction, portions of them advance in other directions.

The most common form of dune at Miller (an excellent place to observe dune formation) is one where the long axis runs from northeast to southwest across the prevailing winds. The sharp, steep side of this dune, it will be noticed, is undercut and worn away by the wind, and the sands stream out to the leeward.

Masses of roots, trees, and the sands of the dune itself, fall from time to time as the wind undercuts them. The undercut is death to the dune, although the vegetation on its sides may protect it for a time. The crested dunes with overhanging peaks are so acted upon by the wind currents that at times they are spoken of as smoking dunes, the moving sand in the air presenting that resemblance.

The sculptural form or shape of dunes is largely

Turret dunes near Dune Park. This unconventional form is unique. Binding grasses protect each crest formation. The photograph, which has been much admired, was taken in winter by Dr. A. L. Van Horn.

dependent upon prevailing winds, but somewhat upon locality. An isolated sand dune is usually conical or shaped like a sugar loaf. These cone-shaped sands are known to have maintained their forms for long periods of time. Their heights remain practically the same, although the top is being blown away all the time, the explanation being that while the apex is greatly exposed, the sands at the base are more so, and the same winds that lower the peak deepen the base. These, sometimes called the pyramid dunes, show greater changes on their sharp, sloping windward sides by striations cut by the wind, but the tentlike form always remains, and the leeward side shows a more gentle windswept slope. At Dune Park these pyramid dunes are very attractive. Possibly they are greater in number here than elsewhere along the lake shore.

The sand dunes, generally speaking, seldom increase in elevation. The general contour remains about the same and retains about the same surface. Their texture in each locality is dependent upon the nature of the rocks from which the dune sand was originally derived. Locality, type of prevailing winds, protection offered, etc., are the causes governing the shape of the dunes, while type of sand and accumulated moisture are the factors that make and keep the form and retain the dunes from marked changes, except as I have elsewhere mentioned.

This sand plateau was formed by the crest of a high dune having been cut through by the wind. It is really an active dune, traveling to the south and east. The tracks in the foreground show its mobile character. The forest to the left is in danger of being submerged. In the right foreground there may be observed a few specimens of the bag-seed plant, a small, hardy shrub which grows very rapidly on these seemingly impossible places.

To those who would become interested critically in dune formation, the sandy beach offers a true counterpart of a sand dune, in the wind ripples along the shore, where all types of dunes are to be found. The crests of the ripples, the undercuts, the location of the coarse and fine sands, all are in place, and a careful study will explain how dunes are formed and what changes them. The ripples are real dunes in miniature.

A furious gale swept over Lake Michigan the night of November 23, 1916. Storm signals in every port warned all shipping to remain within safe harbors. The next day the gale blew itself out, ending in a cold and pelting rain. The following day was clear. The third day the wind had shifted from the northwest to the south and was agreeably warm. In the dunes, the wind had driven the waves high on the summer beach, and in a night had destroyed it—not a wind ripple could be found. Debris had been driven a hundred or two hundred feet inland. The sand everywhere near the lake was pounded smooth by the fierce rain and angry waves.

November 25, 1916, was a rare day in the dunes for studies of dune formation. There was a crust on the sand formed by the rain and the freezing cold, and one could walk easily on the surface. The fresh wind from the south changed to a gale, and during the after-

Little by little this striking profile has been cut by the sweep of many gales. The tangle of roots shows the strength of the winds and the exposed strata indicate different degrees of hardness of the layers of sand. The dune is an old one.

The fallen timber, the standing dead trees, the tangle of broken branches, tell the story of the struggle with the fierce gales. The ripple marks in the sand to the right of the picture denote by their width and irregularity that the wind which caused them had the force of a tornado.

noon a wind storm of the velocity of sixty, seventy, and even eighty miles an hour was recorded. The summer drifts, that had deepened by lazy accumulations, were energized by this violent wind storm, and as the surface of the sand was smooth, the scour of the wind tore loose the exposed parts, especially the crests of the dunes, and with long sweeps across the amphitheaters the blinding sands flew to the northward over all obstacles, up the steep sides, and spread fanlike high in the air in clouds, reminding me of the descriptions I read somewhere of wind storms in the great deserts or, to use a more familiar parallel, like smoke blowing from a chimney top. Time and again during the afternoon we tried to photograph the moving sands that fashioned the smoking dunes, each time recording a failure. These sands were plainly seen all day at the crest of every elevation, some in faint stringlike running clouds, while others, flaring high, sent the sands straight upward, which, in turn, came down in graceful curves far into deep troughs below.

The afternoon was not only a rare day for views, but the topography of many dunes changed from their aspect of the previous summer to forms strangely and startlingly new. Trees three feet high at noon were completely buried at nightfall. The reversal of the traveling dunes from their line of march to the south back to the north was apparent during the day, but marvelously so the following days.

No better evidence of the severity of this gale could be presented than the shore at the eastern end of Sheridan Beach, near Michigan City. For a mile it was strewn with flat rocks torn loose from the bottom of the lake and cast up by the thrust of the waves and left upon the beach, which during the summer had been a stranger to a stone larger than a pebble. Tons of these rocks, some with smooth and round edges, showing the unrest that had attended them in the water, and others with cutting edges, were cast in utter confusion upon the shore to the very base line of the sand cliffs themselves.

THE DUNE SANDS

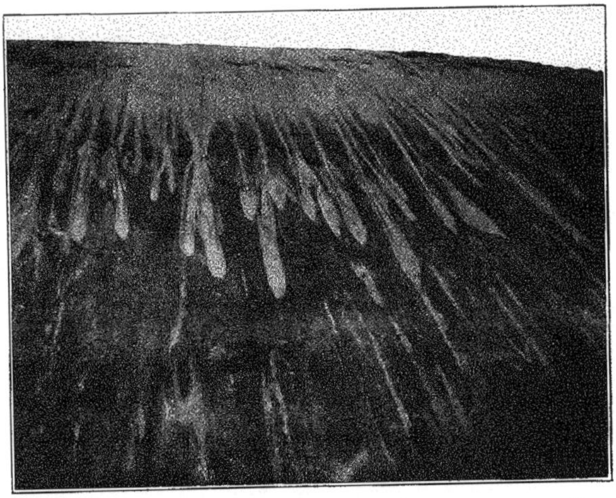

A curious rain effect on the lee side of a traveling dune.

CHAPTER III

THE DUNE SANDS

DUNE sands are very fine sands. Sand hills may form anywhere, but the typical dune sand is purely silicious. The unit for description is not the sand pile, but the tiny bits of silica of which the pile is composed. The flint or silica particles are unattached; they do not adhere, but are separate, incoherent particles, always in a state of unstable equilibrium. They must have had some beginning. Geologists tell us that the topography of a region is always significant of its history.

The sands of a dune country always are imposed upon some basic formation, usually gravels, and the gravels are superimposed upon clay or rock foundations; and, scientists tell us, the rocks, clays, gravels, and sands are associated historically with water. Sands and gravels go to make up the beach lines either of the past or the present, and water placed them where they are found. Dune sands were primarily associated with glacial waters in the same way that moraines are associated with glacial ice. The way the glaciers prepare the fine sands that eventually go to the making of sand dunes is as follows:

When the bottoms of existing glaciers are exposed to view, the lower part of the ice has been noticed to be thickly set with a quantity of earthy, sandy, and stony material of all grades of coarseness and fineness. With

The dune sand is peculiarly soft and pleasing.

these materials imbedded in its base, the glacial ice moves forward, resting heavily upon the surface of its underlying rock mass. The grinding action between the stony matter in the bottom of the ice and the rock bed over which it moves results in the atomic rupture of the silicious surfaces. Sands accrue from these glacial grindings. The glaciers move slowly and with irre-

This picture illustrates a dune in embryo. The scene is the middle beach and the sand is being held by the grasses and shrubs. The plant on the left is the sand cherry; that on the right, the sand reed or *ammophilia,* both typical dune formers and dune builders. Here also are the "singing sands." Photograph taken near the Waverly Beach.

Wind scour of the side of a big blowout, showing the various color layers of sand. The stains are from iron and manganese. The blow-sands may cover these strata in an hour. The morning's rain brought the sand from above in irregular deposits.

sistible force, cutting, crushing, and dislodging; first forward, then receding, reducing to particles every thing that is caught between the upper and nether millstone of this enormous field. The earthy and stony materials, remains of former ages, are washed with the glacial waters, ground by the glacial ice, and slowly reduced to mutable forms, some of these being of infinitesimal size, and when the ponderous glacier melts and recedes, the products of these grindings are left as deposits to be carried away by water to become the shore line of sea or lake.

The south shore of Lake Michigan dates from the close of the glacial age, the number of centuries ago being immaterial. The shore line of the lake was once one long sweeping dune crescent from Winnetka, Illinois, to Michigan City, Indiana, a distance of about seventy-four miles. Wilmette, Graceland, Lincoln Park, Windsor Park, West Hammond, Thornton, Homewood, Glenwood, have had sand dunes similar in their formation to the dunes in western Indiana. Civilization is death to the dune, and buildings, boulevard improvements, ditching, drainage, and permanent barriers have destroyed all but the remnants of former well-known sand dunes on the Illinois side of the state line.

The sand in the dunes is peculiarly soft and pleasing to the touch. It has a charm of association of its own, even if it does fill your shoes and your pockets. It does not feel "gritty," being more like fine flour. It is clean and pliant, and one soon becomes friendly with it, it is so easily handled, so familiarly patted and moulded at one's own sweet will.

Long windrows of pure white drifting sands.

The sand grains vary much in color. The whole crescent shape of the Dune Land is bent more in the center than on either arm, and so in a macroscopic view the range of sands always have changing colors. Some days they appear as white or brilliant gray, at others—according to the sunlight or cloud coverings—they are yellow or pink or purple or golden. In the sunset or twilight hours, they show many more colors.

The moisture in the dunes play some important parts in the various colorings. The same sands are very dark when they are very wet, and very yellow and glistening when they are very dry. The beach sands are angular, those high on the dunes, smooth, shiny, and refractive, and with many colorings. As a rule, all sand grains have a more or less round appearance, due to attrition and to weathering.

Under the microscope, the sand granule appears glistening, with edges smoothed or polished, and often the grains will resemble an uncut gem. They will reveal unsuspected quartz: pink, black, yellow, and garnet colorings.

Sand grains vary in color because the parent quartz or granite rocks have many colors. Some siliceous formations are stained by contact with mineral-bearing rock, or they may have an oxidizing mineral of their own. Sometimes the color of sands that are stratified partakes of the substances that have stained them from above, and the colorings are due to the presence of foreign matter or silt stains.

Bits of white glass or old bottles that have been tossed upon these sands and are partially buried will be found in a short time to be colored a beautiful tint of purple, not unlike Bohemian glassware or a tiny tube that has once held radium. The coloring has been brought about by the action of the manganese in the sand. Grains of sand have both a long and a short diameter, as the microscope will reveal, and they are not as round as they appear to the unaided eye. The grains also vary in size as determined by the various different meshes or measures. Mesh is a technical term; the wire screen of a sieve having a certain size or mesh, the distance of the wires apart determines the fineness or coarseness of the mesh. A sieve with a hundred mesh means that each division is one one-hundredth of an inch; in a five-hundred mesh, each opening is one five-hundredth of an inch. The finest sands are, of course, measured by the finest meshes.

One might easily overlook the polish on each grain of sand, but it is there—two kinds of polish, to be exact. One is due to glacial grindings and water erosions, and the other to the rolling and sliding of the dry sands upon each other. A pocket microscope will show many sand wonders.

It may be well here to define the term sand scientifically. It is as follows: The term sand is applied to particles of stone whose grains measure under one millimeter in diameter—a millimeter equals .03937 of an inch. If the soil has a coarseness of one or two millimeters it is called grits, and when it is above two millimeters and not over six millimeters, it is called gravel.

I am indebted to Pehr Olsson-Seffer for the follow-

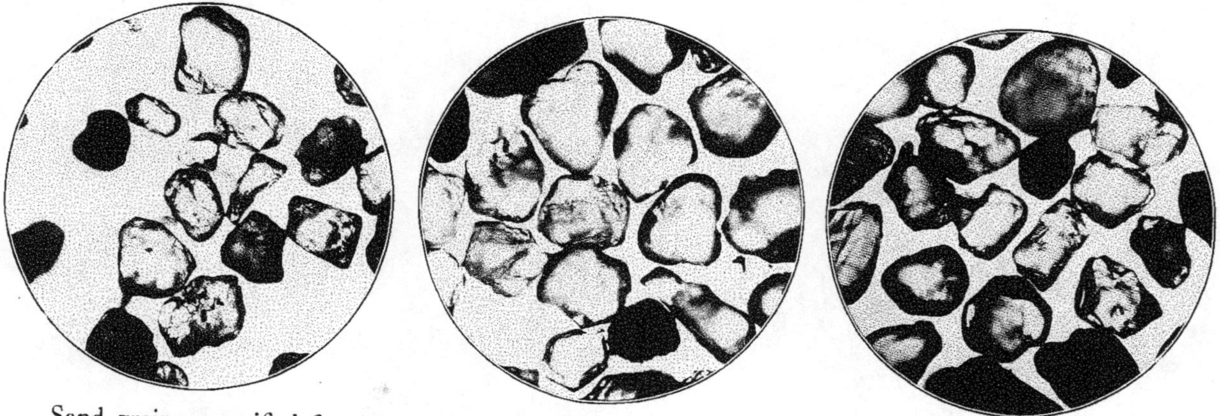

Sand grains magnified five hundred times. The average size of the grains was one-fiftieth of an inch in the longest diameter. Ninety per cent were transparent under the microscope with mostly smooth edges, the parent rock being granite. The grains were taken from the author's shoes after a day's tramp.

A small but typical sand dune. Notice the beach gravels in the foreground, the sand ripples, the fine blow-sand, the furrows which the wind has cut in the exposed slope and the debris of destroyed vegetation. A bleak spot in Dune Land near Carr's Beach.

ing list of measurements. They are instructive in a descriptive way, and belong essentially to dune sands.[1]

	Diameter of grains in millimeters from		
Fine dust or silt	0.02	to	0.03
Medium dust or silt	0.03	to	0.05
Coarse dust or silt	0.05	to	0.1
Finest sand	0.1	to	0.2
Fine sand	0.2	to	0.3
Medium sand	0.3	to	0.5
Coarse sand	0.5	to	1.
Grits	1.	to	2.
Gravel	2.	to	4.
Coarse gravel	4.	to	6.
Pebbles	6.	to	10.
Coarse pebbles	10.	to	20.
Shingle	20.	to	50.
Stones	50.	to	250.
Boulders	250.	and	upwards

All these products are presented to the shore line by Neptunian agencies. The transporting power of water contributes to all land building. Common sand is 2,100 times heavier than dry air, and is but 2.5 to 2.7 heavier than water.

This great division and subdivision of silicious matter, as we have already noted, has once been associated with the action of water. The way these parts are again rearranged is to make both beach and dune formation readily understood. We recall, too, that in the beach lines there is a winter beach and a summer

[1] *Genesis and Development of Sand Formations on Marine Coasts.*

beach. The former is fashioned by storms and gales driving higher upon the shore the silicious sands, gravel, pebbles, and stones; while the summer beach has a rhythmical accumulation of sands that are fine and still finer. I offer two illustrations and a few words of description to make this very plain.

The month of May, 1917, was a very stormy month in the Dune Country; there were many changes because of the strong gales. May 13, 1917 was a windy day. It had blown steadily for three days from the north. The waves were breaking on the sand bars far out in the lake, and came pounding in great force upon the sandy shore. The water was laden with stones, pebbles, and sand. I watched the waves for a half-day, and practically saw the stones in the foreground in the illustration on the opposite page, washed upon the shore and distributed in this irregular fashion. While the water was sand laden, very little was deposited. It went with the backwash into the lake. The stones were rolled up from the bottom of the lake, and were deposited by the thrust of the waves, the retreating water failed to dislodge them and they remained as a new floor for the high beach line. The shifting sands from the summer beach will one day cover this deposit.

The illustration on page 73 is of the low, or summer beach, with a summer sea, on a summer day.

The wave lines of sand left by the retreating waves can be traced by the irregular bits of colored sand on the smooth shore. The very fine sands are carried on the crest of the incoming wave. The very top portion of the water, this time gentle and lisping, rather than pounding and roaring, carries a tiny bit of fine sand, and in the final drive upon the shore deposits a thin,

After the storm.

broken line of sand, and then the wave retreats. These wave lines are always irregular, owing to the forces that act upon the water—such as, the bottom of the lake, varying breezes, deep and shallow shores, etc. If the waves should freshen and come in as breakers or combers, then the action on the shore line would be to destroy it. When the water is stirred to the bottom of the lake, the stones and rocks are dislodged and floated inland, but in the peaceful break of a wave upon a summer beach the trace of fine sand that is left will in time dry out, the wind will stir it, and the sand will have a final resting place as part of a great dune. These tiny strips of variously colored sands, looking so useless, are, in reality, the units out of which dunes are formed. In fact the secret of dune building is here revealed, being dependent upon the size of these silica particles and the way they are acted upon, first by the action of the water and then by the wind.

As a wind blowing upon snow causes it to drift and also to ripple its surface, or as wind striking the surface of water at a given angle stirs it into a ripple, which is deflected by some obstruction, so the wind changes the sand grains according to their sizes. The first showing of a dune formation is the wind ripple. This can occur only when the grains of sand are of different sizes.

The rippling is aeolian, and it happens where the sand grains are shoved along, tumbled over upon each other rather than raised by gusty winds. The wind action on the sand is to form a crest having the coarsest sand at the top and the fine sand at the bottom, and between the two crests the wave ripple has all the elements of formation that a great dune has.

In his recent book, *Waves of Sand and Snow,* Vance Cornish discusses the rate of advance in wandering dunes and the size of the sand grains. His studies cover

Wave marks on the lower beach. The sand comes to the shore on the top of the wave, which, when it recedes, leaves a tiny elevation of fine sand. It is the presence of black grains of sand which makes the lines of deposit so clear. If not disturbed these threads of colored sands become the first unit in dune building.

Detritus that helps make the shore lines. These white pine trees were felled for lumber about seventy-five years ago. The stumps are well preserved, having been covered with sand the greater part of the time stated.

a world-wide observation of dunes, and the measurements are many. He finds that the size of the grain of sand in the wandering dune, that is, the average traveling dune, varies anywhere from one forty-eighth to one ninety-second of an inch in diameter, very often the latter, and in a wind averaging twenty-five miles per hour, the rate of advance of sand grains varies from twenty-five to forty-two one-hundredths of an inch per minute.

High wind velocity is necessary to the distribution of the various sands. The transportation of the sand from its wave deposit to the crest of the dune is a mysterious process. It takes a strong wind to stir the dust in the ordinary roadside, and yet the specific gravity of dust is much less than that of sand. Ordinary winds simply blow over the ripples and leave them unchanged.

The lifting power of the wind is very slight. When it moves the sand at all it is in a sliding and not a lifting manner. The motion most often noted is a twisting or stirring of the upper surfaces, causing trituration or grinding movements, producing erosion of surfaces. The grains thus diminish in size, and the parts lost by each grain go to make the finer subdivisions. A fine sand dust when dried is blown by this wind power, and this is the first of the beach sands to be fashioned into the dune shape. Professor Olsson-Seffer offers an opinion that the winds that are the chief power in the aeolian geological work are those of storm velocity; which means a velocity from fifty to one hundred kilometers per hour—a kilometer being about five-eighths of a mile. Dune changes, then, are dependent upon winds averaging sixty miles per hour or more.

Wind velocities are measured usually at some distance above the ground, but the dune sand is moved only by the very lowest strata in the atmosphere. This lowest layer is increased at a very slow rate, with an increase in the speed of the layers next above. The velocity in the layer next to the surface of the ground probably never reaches three miles per hour. It is this fine layer which alone comes in contact with the resting sand and first causes it to stir.[1]

The transporting power of the wind varies more nearly in approximation to its erosive force than to its lifting force. The conclusion of numerous analyses is that the wind forms drifts of grains which measure one-half to one-eighth of a millimeter in diameter. The lee drift of sand is found to be a little finer than the dune sand proper. The grains have been lifted a little higher, and that is the reason they have been carried a little farther.

When the wind is blowing a steady gale, the amount of sand actually moving is not large. Much more sand is changed when the wind is fitful or gusty. The ordi-

[1] J. A. Udden, *The Mechanical Composition of Wind Deposits*, p. 24. Augustana Book Concern, Rock Island, Ills.

A sand dune of the amphitheater or horseshoe type. The scooping-out process has not been rapid. The lee sands may be seen in the background. The sides show the destruction of vegetation and the uncovering of dead tree trunks. The grasses in the center are beginning to take root again.

The lee of a traveling dune, showing the angle of soft sand decline by gravity. The pine tree to the left has been covered to the extent of thirty feet from its former base. The advance of this dune was very rapid. Photograph taken near Dune Park.

nary sands which go to make the bulk of the dunes are not carried very far—a range, probably, only of about two hundred feet. The fine sand is transported a few hundred feet farther, but the atmospheric sand dust,

Sometimes the sand attacks the forest from two sides.

the finest subdivision by erosions, is carried long distances. Sometimes it is possible to notice an almost indefinite suspension of the finest sand dust in the changeable currents of the atmosphere. These tiny particles are carried several miles.

Coarse gravels are common in the dunes, but boulders are rare visitors. Occasionally a granite boulder is dug out of its long resting place. It is a stranger to the parts, and must have wandered far from home, for those that have been found here belong to formations that are known only to Canada.

To conclude this chapter I will give an example of how the sand dunes may be studied by the application of such analyses. I copy from Professor Udden's book a table showing the mechanical composition of dune sand near Michigan City, Indiana. Dunes at this location are strangely beautiful and very active. They are very high and from the lake side are not unlike chalk cliffs in appearance. The shore line is abrupt, precipitous, showing mixtures of clay, gravel, detritus from materials carried in from the waves. This illustration might also be called, in its final analysis, the type of the Indiana Sand Dunes.

The table reads as follows, in terms of millimeters:

LENGTH OF DIAMETER THE SAND	UPPER REAR SLOPE OF DUNE	DUNE SAND	TYPICAL DUNE SAND	NEAR CREST DUNE	CREST OF A DUNE
8 to 4 mm.....	.02	.1	...	...	...
4 to 2 mm.....	4.1	1.1	...	12.4	...
2 to 1 mm.....	7.4	8.2	.9	6.4	.4
1 to $\frac{1}{2}$ mm.....	14.1	25.1	20.7	13.8	6.0
$\frac{1}{2}$ to $\frac{1}{4}$ mm.....	26.7	64.0	76.8	65.2	92.6
$\frac{1}{4}$ to $\frac{1}{8}$ mm.....	45.3	1.2	1.0	1.8	.6
$\frac{1}{8}$ to $\frac{1}{16}$ mm.....	.5	.2	...	...	...
$\frac{1}{16}$ to $\frac{1}{32}$ mm.....	...		..	...	

With this table before us we learn that the upper rear slope of this dune has one-half of its sand in three sizes; that the average size of the dune sand is composed of sand from twenty-five to sixty-four per cent, varying from one-half to one-fourth of a millimeter; that the typical dune sand has seventy-six per cent of sand of one-half to one-fourth millimeter, and near the crest has both coarse and typical fine sands; and at the very crest, ninety-two per cent sand, or sand one-half to one-quarter of a millimeter.

Again, the wandering sands are traced by their sizes. This table, compiled by the same author, tells of the distance the sands may be wind-borne—

Fine sand, $\frac{1}{2}$ to $\frac{1}{8}$ mm., less than a mile.
Very fine sand, $\frac{1}{8}$ to $\frac{1}{16}$ mm., a few miles.
Coarse dust, $\frac{1}{16}$ to $\frac{1}{32}$ mm., two hundred miles.
Medium dust, $\frac{1}{32}$ to $\frac{1}{64}$ mm., one thousand miles.
Fine dust, $\frac{1}{64}$ mm. and less, around the globe.

The erosion of quartz and igneous rocks by waves and currents have not been considered here, as the operation of dune building is with the dry sand, for as long as the sand is kept wet by the spray or by capillary attraction, it is not disturbed by air currents.

Starved out.

OLD AND NEW BEACH LINES

Century-old oak trees engulfed by a traveling dune. The footprints in the foreground show the loose formation of the sand. Photograph taken near Michigan City.

CHAPTER IV

OLD AND NEW BEACH LINES

THE waters of the great inland sea known as Lake Michigan extend northward from the Dune Country more than three hundred miles. The south end of this great lake is the beach side of the Indiana Dune Country. The water is shallow for a long distance out from the shore line. The bottom of the lake has ridges of sand, the sand bars, which may be seen from the crests of the high dunes. The prevailing winds follow the isothermal lines and are from the north and northwest, varying from north to northeast, so that the Dune Country is in the direct line of all prevailing storms and winds. In winter the storms pack up the ice in great drifts along the shore, but the ice does not reach the dunes or assist directly in dune formation or destruction. The ice does, however, pack and grind the sands to a finer mesh.

Wind storms often sweep the shore lines in a savage rage. For centuries the waves have been doing the same as now. Many of the old beach lines have been utterly destroyed and new beaches are constantly being made. These beach landmarks are interpreters of the past and present. They record the histories of the Dune Lands, and the story of the dunes would not be intelligible or complete without the story of the beaches.

At the south end of Lake Michigan the new sand and gravel beaches are perfectly defined and permit an easy survey. The water line is paralleled by a land elevation which is an almost exact counterpart, in nearly the same crescent shape. At some parts it is quite precipitous and it varies in distances from the lake from a mile to fifteen miles to the west and to the south of the present lake line. This variously elevated land is known as the Valparaiso Moraine—a moraine being a mass of earth, stones, and boulders deposited by glacial action. The name has been given this land formation because the city of Valparaiso, Indiana, is located upon it. The Indiana west-state-line boundary is the east line of the state of Illinois. This division is purely legal, but helps make for accuracy in the description of the various beach lines.

The Valparaiso Moraine is traceable from a point twenty-five miles to the north of Milwaukee, Wisconsin. It has several extensive breaks in it, but the inner or eastern margin of the moraine is strong in Illinois at Elmhurst, LaGrange, Palos Springs, Homewood, and Glenwood. At Dyer, Indiana, the moraine is fifteen miles south of Lake Michigan, and at Michigan City, Indiana, it is three miles to the south of

A low harried dune. Its windward face is exposed to cutting winds. The sand layers here are packed hard and show various colorings. The crest of the dune is being held by the vegetation and the base is accumulating protecting sand.

that city. A large relief map of the boundaries of Lake Michigan and the northern line of the Valparaiso Moraine, can be consulted, if the reader chooses, in The Field Museum, Chicago, alcove No. 106.

The beginnings of the Dune Country must be studied from land markings and local geological maps which show three distinct shore lines, and these shore lines have left their evidences on the side walls of the moraine. The exact lake lines, corresponding to land lines, are impossible to obtain. They are inconstant and were of ice. The old beach lines are not so easily traced, but there is sufficient data to describe three separate and distinct beach formations that are exposed on the plain between Lake Michigan as the boundary to the north and the moraine to the south. They represent different ages for the present lands, and explain, to some extent at least, why the variations and location of sands are as they are today.

Lake Michigan has an elevation of 581 feet above sea level at tide water at New York, and the Valparaiso Moraine is 600 to 700 feet elevation at Lockport, Lanes' Island, and Blue Island, in Illinois. It is estimated that at one time the moraine elevation was more than 700 feet.

A large body of glacial water known as Lake Chicago, existed at one time; conjectured to have been from

20,000 to 100,000 years ago. The exact dimensions of this lake can never be known, but generally speaking, the first boundaries of Lake Chicago are approximately these: to the northwest and west and south, the moraine, while the east and north boundary was marginal to the great ice field. The water of Lake Chicago rose to a point sixty feet above the present level of Lake Michigan, or 640 feet above tide water at New York, and was twenty to sixty feet below the top of the Valparaiso Moraine at various points.

The first beach line created by Lake Chicago is well defined at Glenwood, Illinois, and also found, but broken somewhat, at Dyer, Indiana, and eastward from Dyer. It extended westward to The Sag, in Illinois, northward to Austin, Oak Park, Maywood, and Norwood Park to Waukegan, Illinois, to some narrow point near Michigan City, Indiana.

The first drain or outlet of the water of Lake Chicago occurred at The Sag, where the Des Plaines Valley helped to carry the water south and southwest to the valley of the Illinois River.

Geologists tell us that the evidences do not point to a waterfall at The Sag, but there was a series of rapids with an immense volume of water rushing to the south and west, and this outlet must have been comparable to that now discharged through the Niagara River. The lowering of this glacial lake left a beach line known to geologists as the Glenwood Beach.[1]

The almost complete drainage of Lake Chicago, which happened later through some unknown outlet, had a great deal to do with the sedimentary deposits found in the immediate vicinity west and south of the present boundary of Lake Michigan, which is known as the Indiana and Chicago plain.

The topography of the Chicago plain is very uneven; running to the south and east, is a series of mounds, elevations, islands, and depressions forming lakes, rivers, and very extensive marshes and lowlands. Sometimes the plains dry out, but heavy rains and the spring and fall freshets cover vast areas with a shallow sheet of water, often stagnant, for it is never really drained. This is believed to have once been the bed of Lake Chicago.

Subsequently, Lake Chicago filled again to a point forty feet above the present level of Lake Michigan, and twenty feet below the Glenwood Beach line. This is known as the Calumet Stage, and is evidenced by the beach lines, five miles or more south of Miller, Indiana. Other markings at various points are also in evidence.

The beaches along the Calumet Stage indicate that the general configuration of the country must have been

[1] For a full description, see *The Geography of Chicago and Its Environs*, Bulletin No. 1, p. 51. The Chicago Geographic Society. R. D. Salisbury and William C. Alden.

The life of the pines is strenuous, and years of struggle with the elements make them sometimes assume grotesque shapes. The picture shows the ravages storms make in the tops and how the wind breaks and twists the limbs.

much the same as now. Critically examined, there are no less than ninety sedimentary elevations from three to ten feet in height above the plain. They constituted the high spots in the bottom of the lake. The uneven shore lines, consisting of small dunes of very irregular gravel and sand deposits, can be imperfectly followed at this very late time. These markings all belong to the Calumet Stage. In due course of time the lake again emptied. By reason of this stage of land making, the drainage of Lake Chicago was diverted from The Sag; the currents were reversed, the Calumet and the Chicago rivers were formed. The waters of the Calumet were never very deep. They flowed easterly emptying into Lake Michigan at Miller, and the waters of the Chicago River drained to the north, finding an outlet a mile or more to the north of its present outlet into Lake Michigan. The same influences changed the slightly undulating surfaces of the Chicago plain to a plain exceptionally flat which is practically the topographical appearance today.

Once more Lake Chicago rose to a point sixteen to twenty feet above the present level of Lake Michigan. Again some mighty convulsion of nature must have happened, for the marginal ice-field boundary to the east afterwards turned its glacial waters into what became Lake Michigan. The most distinct beach markings to be found are at Tolleston, Indiana, about five miles west of Miller and the marginal lake water was then respectively three, two, and one miles to the north of Tolleston, Miller, and Dune Park.

The sand dunes of Rose Hill, Graceland, and Windsor Park, Illinois, and of Hammond, Tolleston, Miller, Dune Park, Tremont, and Michigan City, Indiana, came into existence at the same time as the Tolleston Stage of shore or beach lines.

With the final disappearance of the waters of Lake Chicago, various depressions were left—such as are now called Hyde Lake, Calumet Lake, Wolf Lake, Lake George, and the two rivers, the Grand Calumet and the Little Calumet—each with a shallow depth and currents so lazy that their outlets can scarcely be traced. Lake Chicago left also some elevated or island areas—Mount Forest Island, Stony Island, Blue Island, and many lesser ones.

The ice covered by sands wind-driven from the land.

Prickly Pear Cactus (*opuntia vulgaris*). A thick-leaved prickly species of cactus very plentiful in the dunes. It grows rapidly, bears delicate yellow flowers and produces fruit. The photograph was taken in the winter.

THE DUNE DESERTS

The dead tree trunks here are the remains of a former forest, now being uncovered by the wind. Such is dune life. The sand engulfs living trees and holds them in its embrace for years. Then the wind reverses the process and under its influence the sand is made to disgorge the lifeless trunks. Photograph taken near the Chesterton road.

CHAPTER V

THE DUNE DESERTS

"IT IS not necessary to go to Africa to study desert phenomena, for Indiana has a genuine desert, though a small one." Thus writes Hu Maxwell, and he gives several reasons in support of the assertion. The idea is interesting, and it is but a short trip from dune to desert in this instance. Describing a desert, he says:

> Every change in the wind produces changes in the contour of the barren tracts. Sharp-crested ridges become rounded; steep slopes are modified; peaks quickly shoot up where none were before. When the surface is dry and the winds are stormy, a surprising quantity of sand is on the move, but it seldom rises in the air more than a few inches, and is apt to escape observation unless a watch is kept when it passes over a sharp ridge. To leeward of a crest over which it has passed, it fills the air like a cloud, then drops to the surface and slides down the slope under the stress of gravity, until friction brings it to rest.[1]

A desert, then, is a curious region of constantly changing sand. A desert may seem to be a result of accident or neglect or something to excuse for being at all; but really a dune or a desert is a most orderly sequence. It is a certain tract of land that was fashioned in the beginning exactly as it now is.

[1] *Scientific American Supplement*, No. 2076, Oct. 16, 1915.

The desert with its changeless yet ever changing plain never fails to stimulate the imagination. Its sands, exposed to the winds, make strange outlines with the horizon. The low brush, now stunted and burnt and dwarfed, gives place around the next bend to brush, tall, thrifty, and of luxurious growth. It is apparently the same old desert, but the inhabitants and others who tread its sands know that it is never the same for an hour, a day, or a trip. Incessant change is the law of the desert.

Without vegetation, the desert is a boundless sea of sand, such as for instance the Great Sahara Desert, of which a noted traveler and lecturer gives a vivid word-picture.

> Hundreds of miles to the westward stretches the vast Sahara, scattering its first golden sands at the very base of the pyramids. It is an awful sight from its dreary immensity. With its rolling waves of sand it seems a petrified ocean suddenly transformed from a state of activity into one of eternal rest . . . . in truth this vast Sahara is an ocean of sand. It has the same succession of limitless horizons and the same dreary monotony . . . . finally, like the ocean, it is often lashed by storms which sweep it with resistless force, raising its tawny waves to blind, overwhelm and suffocate the wretched traveler who may encounter them, until he falls, coffined only in the shroud of sand woven around him by the pitiless storm-king.[1]

The drifting, driving sand in the little deserts of the dunes has no danger zone, nor do we encounter miles of unconfined sands, or fearful sand storms, but

[1] *John L. Stoddard's Lectures,* Vol. II, p. 326.

Abatis made by the blow-sand. The roots of a white pine were uncovered by the sand dredge at Dune Park. Nature's sand blast has sharpened the roots to bayonet points and polished them and the exposed parts of the trunk to a piano finish.

the sand itself has a cutting and abrading power. Take the case of some fallen tree lying athwart the path of the prevailing winds; if the roots are turned upward, the trunk with its stubs of branches remains exposed to the constant sand blast. In the lee there may be a streamer of exceedingly fine sand against some obstructing root, which has become covered, otherwise the whole tree lies continually exposed with its wood in a fine state of preservation, though it was a century ago when its life period was at its greatest. The savagery of the attacking sand where the storm is fiercest, not only has stripped the roots and trunk clean, but it has polished all exposed surfaces smooth. The stubs of branches remaining fast to the trunk are whittled and sharpened to bayonet points and the smaller roots are made into spears. The repeated polishing makes it impossible for the sand to lodge over, or even underneath the trunk or around the roots, and aids the wood to resist decay.

The study of tree life in the desert dune is of real interest. The needle-leaf trees, pines, and cedars, which appear to thrive elsewhere in adverse places, are precarious and do not grow abundantly in the sands, but the broad-leaf trees such as cottonwood and basswood, seem to find the dunes a congenial habitat, thriving on the sides of some bald-topped dunes and on the very summit of others one hundred feet or more high. The broad-leaved trees may be found anywhere in the dunes, even in the swails and marshes. The uplands are their real home, but apparently they are able to adapt themselves to the barren sand, sending out enormous roots scouting in all directions for the moisture they demand.

The passing of a once stately pine.

At various points in the high dunes, two or three forests, one above another, were successively and completely buried with many of the tree trunks standing. The trees usually die after the sand drift has accumulated a few feet high around their bases. Once completely covered by the sandy surface, decay ceases and the stubs remain embedded in sand until the time comes when the burial mound is blown away, exposing the tiers of stumps and of branches and of roots and of barkless trees, some fallen, some standing.

Red cedar stumps are occasionally uncovered in these sand tombs, also cottonwoods, basswoods, white elms, and several oaks, principally the jack and burr oak. Splendid white giant pines and the more common jack pines existed in the dunes in the past centuries. Their remains, together with their vegetative peculiarity, prove their antiquity.

The uncovering of the roots of standing trees, as the wind scoops the sand out from around them in many places, reveals the humus line, as it existed at one time. A long tap root that keeps going down and down may frequently be seen, still holding the giant of a former day in an expectant balance. When the wind has secured an entering eddy, forcing itself in and the sand out, the doom of the tree is sealed—it is simply a matter of time. Any moment may see its fall. It is the breath of death, the first withering, the beginning of the end, the law of tree life in the entire desert zone.

Should the remains of a former period of vegetation resist the attacking winds, some weak spot will be found, for the spirit of the wind is as persistent as is time itself, and the scooping-out process rather than a blowing-down process may prevail. Eventually the foundations are undermined, and the trunks of thrown trees are let down lower and lower until they accumulate in the bottom of some ravine or on a barren slope,

This cedar tree, once upright but now prostrate and deformed, still keeps up its struggle for existence.

perhaps fifty or a hundred feet from where they once grew.

The chief matter of interest connected with the Indiana Desert, says Hu Maxwell, is not so much what grew here in the past and was buried by drifting sands, as what now grows and how it obtains a start and how it is able to maintain its foothold. This refers to trees and other perennials, the very existence of which presupposes fixedness and stability of soil. Their advent into life, their growth, death, burial, and excavation, may be studied side by side, and often in very small areas. He speaks of an incident which shows the uncertainty of plant life among the dunes.

Two visits were made to the locality, a period of fourteen days intervening. At the first visit a certain plot of an acre or more was observed on a gentle slope of an old dune. The place was at rest. There was no indication that a grain of sand had drifted there in fifty years. Young trees from five to twenty feet high covered the tract. The small trees attracted particular attention, because they were so close together and were laden with fruit, which was then green, but would be edible when ripe. One was a wild black cherry (*prunu serotina*); one a nanny berry (*virburnum prunifolium*), and one a service berry (*amelanchier canadensis*). The occasion is rare when these species may be seen growing side by side, and the fact that they were so growing in a place so unkindly as the slope of a sand dune, was sufficient to attract interest to that phase of ecology. . . . At the return visit two weeks later, a somewhat difficult climb was made across a sliding slope in order to see the same trio of little trees again. A profound change had taken place. It was at first difficult to recognize the spot. The trees and all their near associates had vanished. The sand had taken a turn in that direction and had buried them and at that moment it was pouring down the newly formed slope and was rapidly burying the young oaks and sassafrasses lower down. In other parts of the desert area, two weeks sufficed to change many features of the topography. The suddenly increased activity was apparently caused by new attacks of the steam shovel on the ridges near the lake, but so far away that the connection between causes and effect was difficult to explain.[1]

---

[1] *Scientific American Supplement*, No. 2076, Oct. 16, 1915.

The wild grape vine grows in great profusion in the dunes and opposes dune destruction. When it has climbed a tree or shrub its topmost branches turn earthward and when the ground is reached every node forms a rootlet in the sand. The roots of the vines travel long distances to obtain moisture.

The one plant more typical of the desert than any other is the cactus. It seems that every desert has its own species, and the unloved plant is found in great quantity in the Indiana Dunes. In fact, it grows rapidly in great patches and is difficult to exterminate. A fragment broken from the roots may be thrown aside and left exposed for days and yet it is sure to grow. The plant bears delicate yellow flowers and produces fruit. The dune species is declared to be similar to the devil tongue, which has made the cactus valley of the Rio Grande famous.

The plant known as *equisetum hyemale,* or horse's-tail, is credited to this region, growing to about a foot in height. Its ancestors, centuries ago, were tall trees, and their ribbed trunks contributed enormously to the formation of coal in the carboniferous period of time. In all things, except size, it looks as its ancestors did so long ago. Variously changed characteristics could be noted in the description of many other plants, due to desert environment.

The wild grape is another hardy dune vine. Its roots travel long distances seeking the water without which it could not live. I traced a single grapevine root for over sixty feet, and then found that it was scarcely half of the distance the root had gone from its stalk. The vine propagates itself where the joints are buried in the sand. The node throws out a new shoot and from the burial the shoot becomes a root. The grapevine through its tendrils holds tenaciously to any upright. Its habit is to climb and to cling. It swings its free branches clear from its supports and they hang in graceful forms until the nodes again touch the earth, when other roots are formed, and thus an added resisting power against a mighty wind is in the making.

The wild rye is one of the grasses very much in evidence in the desert, and it is defiant, too. It grows anywhere, and, if it chooses, exactly in the center of greatest wind-exposed places. Its roots are strong, going in every direction, and almost always in the lee of every clump of wild rye is a battlement of fine sand caught and held in the meshes of the growing stalks.

The roots form such masses that when they are exposed they are often found to be fifteen feet or more in width, and eight or ten or more feet in height, looking like great sprawling mounds, and at a distance resembling haystacks in a snowy field.

Fairy forms of ripe grasses.

THE DUNE HINTERLAND

Head waters of the Grand Calumet.

CHAPTER VI

THE DUNE HINTERLAND

THE sand dunes, in a ridge of varying width, skirt the line of the lake, but just behind them are what may be called the Dune Lands, comprising many miles of picturesque and really beautiful landscapes.

From the car windows of the electric line, which runs through this hinterland, often more than a mile to the south, there is no finer view of the skyline of the dunes fashioned now by forests and again by sands. Every now and then great patches of white sand tower aloft, having for their background the blue of the sky or the scud of clouds. Often tall trees, some dead and black, others living and green, stand on profile points, and one sees them as watchful sentinels guarding their lands. At times the forests hide the sands for many miles, while from the edge of the lowland, tier upon tier of forest trees, many-hued, make a mountain bank of green tops against the blue or gray horizon. The sand ridge itself at times abruptly breaks and seeks the level of the plain, and then again it rises sheer and is phantastically undulating.

If you choose the more leisurely way of walking, which really is the way to see Dune Land, do not fail to stop and admire the scenic points as they burst upon your vision. You may not get very far in a day because such little obstacles as little yellow butterflies by the roadside or a field of cinnamon ferns or a colony of lady slippers or a nest in a tree or in the rail fence, may cause serious blockades to your adjustable timetable. It matters not, however. You have many more such days before you, for it soon becomes apparent to the hiker that here is one of the most interesting places in the world.

One finds many surprises in the lowlands. You may find a tree that leans way over a shelving shore and makes such a shade that you resolve to rest there some day in summer and bring your favorite book and read, but you never do so; if you bring the book, it is neglected. You see, instead, the living story and try to read that. You watch the turtles crawl to their sunny plank or stump or root. Yonder may be a teal duck or a diver; it's their home, and they may wonder why you intrude. Frogs and crawly things announce their disapproval of your presence and disappear. Did you ever watch them and see their cunning, and by waiting see how stealthily they reappeared and watched you? The swamps have ten thousand eyes.

Every trip finds something new, and things grow into importance—how great they may become is dependent upon yourself. The wonder-workers here are

Here a path is left between two traveling dunes. The crests are wide apart and the path is wind-swept its entire length, which is about a hundred and fifty feet. Photograph taken near Dune Park.

the little builders. The tiny home of insect, water bug, or worm in the swamp land may be of exquisite design, beautiful in finish, and marvelous in efficiency.

One day the pussy willows were in full bloom; later I saw them with the long silvery plumes that end

Pussy willows.

in feathery flights like thistle down, and either float on the surface of the pools to a favorable shallow spot, where a new seedling may start, or perchance are blown away and are lost. Another day I stopped in front of the same bunch of willows where some weeks before, when the ice had scarcely freed itself from the roots, I had focused my camera in the center of the bush to get a picture of the very first bud. Four different times I interviewed the same little bunch of willows. I began to know them as I know my speaking friends.

Suppose you are starting your day in the dunes at

The irresistible advance of the dune.

Michigan City. The time, we will say, is the spring, near the middle of May. You are standing halfway to the crest of one of the finest dunes in all Dune Land, looking out on Lake Michigan and then down the sharp windward face of the dune to the gravel beach strewn with debris from the wind-lashed inland sea at your feet. It is here as your gaze dwells upon the beautiful view that the spirit of the dunes will come to you, and you will begin to understand the lure of the Dune Country.

Climbing to the crown of the dune, the view is that of a typical desert, though a small one, for here are the shifting, billowy sands all about you. Then as you go farther in a walk of thirty minutes, through the soft and yielding sand to the south, the scene has again changed completely. You have arrived in the flower gardens just through the lee sands. The wild flowers at this time of year are forward, possibly they have been better protected by their locality. The wind flowers have come and gone. In one of the valleys the purple lupines, wild flowers of striking beauty, have preëmpted acres of the woodland for their spring floral festival.

Scarcely have you left the fields of the lupine when you find other meadow lands, or oak openings, carpeted with wood violets. The profusion of blossoms on a single plant is a reason for the myriads of bloom in these fields. One does not often have such a chance to see the many-hued violets in such a riot of leafy costumes and alluring shades. Possibly it was here that the thought of naming the violet, the Indiana State Flower, and the oak, the Indiana State Tree, was born. Somewhere such a thing was suggested.

Going westward, the openings show many kinds of

A field of wild lupine. The wild lupine or wild pea delights in sandy hillsides and protected hollows, and blooms profusely in May and June. It is in great quantity at the dunes, and wide expanses of it may be seen in its flowering time.

trees banked among and against the sands. There are many tall pines, of the age of a hundred years and more, which point their tops high in the sky; and there are thickets of jack oaks and of scrub oaks

The storms have uncovered the roots of this tree but it lives on; the roots for the present, having met the new conditions.

and jack pine trees, large and small, some true to type, others stunted and gnarled; while the beeches, the basswood, the ash, the tulip and the wild cherry, together with thickets of larches, aspens, alders, and willows, thinly or thickly growing, with symmetrical forms and tender green, decorate the color sides of the sand hills in a way that challenges instant admiration.

The type of the tree life of the dunes on the lee side is the oak tree. The lee forests are oak forests; not alone a single kind of oak tree, but all the kinds of oak trees. The type of the northern slope is the pine, also of various

In the path of the advancing sands.

kinds. The lee slopes of the high sand hills afford a special protection to tree and plant life, and but a few hundred feet from where on the northern slopes the evergreens denote the struggle against almost arctic exposures, there may be seen to the south on the sides and at the foot of the southern slopes, behind the barriers of sand, a luxuriant forest of beautiful oaks in perfect foliage. The illustration of a protected forest is from a photograph taken in the region of the Mount Tom Range of dunes. The topographic forms show rapid changes here, beginning at a bridge at the entrance to the Prairie Club's farm land. There is a water-soaked

The cottonwoods on the right are taking advantage of the protection given them by the tangle of vines and roots.

humus, a lazy stream with practically no current, no drainage, abrupt sloping elevations, a native spring, elevation after elevation with a background of a range of dunes away above the top of the tallest oaks. Many are the exclamations of surprise at finding the arctic evergreens and the desert oaks, all within a half-mile walk.

Prof. H. C. Cowles speaks of these phenomena as follows:

> We can not divide plants into those of clay, rock, and sand, but must take into account that most plants have a wide range of life so far as soil is concerned, provided the atmospheric conditions are right. The chief exception to this statement seems to be found, not in the original soil, but in the superimposed humus. There are many plants that require humus for their occurrence in nature, but it makes no difference whether the sub-soil is rock, sand, or clay, provided that the humus alone is present in sufficient quantities. It is by reason of this last fact that the mesophytic forests can appear in all conditions in this climate, since the mesophytic forest is associated to a high degree with humus.[1]

The flora of the dunes is very extensive and orchids and cacti are not infrequently very near neighbors. Flowering trees like the wild plum, perfect their blossoms, though often dwarfed, but do not ripen their fruit. For the benefit of those interested in plant and tree life in the dunes I will mention that the directors of the Academy of Science, Chicago, have located in their building in Lincoln Park, on the third floor, an exhaustive study of the flora of Lake and Porter Counties of Indiana. The exhibit is presented in large, colored photographs, every care being taken to depict each peculiarity of every family group. The classification

[1] *The Plant Societies of Chicago and Vicinity*, page 55.

is by showing the flowers of the month they are in blossom.

The directors have also authorized the building of a new exhibit of a typical sand dune, with beach lines, real sands, and detritus, and with accessories that are to be prepared by artists who have already secured

How tree life is destroyed. The wind has effected an entrance to the dune and the typical undercutting of the trees has commenced.

and preserved the entire color scheme of the Dune Lands. The photographs will represent some of the most expert and expensive views modern art can reproduce. The completion is announced to be within a year, the work of building being now in progress.

The trails leading from the Tremont, Chesterton, and Mineral Springs roads, that skirt the sand hills, become more fascinating as the hidden and wild beauty is discovered. A few visits are necessary to begin the acquaintance. One can wander in any direction and be certain of being startled with pleasing discoveries. The whole country here is beautiful. The foliage and vegetation on the lowlands is often perfect in type and attractive in form. Three men, each with a camera, stood in the middle of the road at the Chesterton station late one evening last year. One of them called attention to a lone tree in a nearby meadow. It was a perfect oak tree. Another picked a faultless elm tree and made a type photograph, while the third picked an ideal maple and a tulip tree. One lovely spring day, just around the bend of this road, at the water line of a marsh, I came across an acre of water-covered land, purple with the Japanese iris in perfect bud and flower. The effect was startling and beautiful.

These old wagon roads seem favorite places for teachers with pupil classes. From the universities come grave and learned professors to study Dune Land. Botanists, zoölogists, and ecologists visit here frequently. College societies and student social societies often come, and possibly it matters little if the studies close with a "bacon bat" around a beach fire as the day ends.

Go to Dune Park, which is the next point; spend the entire day along the banks of the Little Calumet River. For a time the dunes are lost to view. Another day, hike down the trunk line railroad, westward from Dune Park, seven miles, the objective point being Miller. You will pass by some habitations that show considerable age but little prosperity.

Miller, which is practically the western boundary of Dune Land, is an old landmark, and was put on the map when the Grand Calumet River once drained its waters into Lake Michigan near this point.

THE LAKES AND SWAMPS OF DUNE LAND

A bank on the shore of Long Lake. Willows have invaded the water and the margins of the lake are rapidly filling in. First stage of land making.

CHAPTER VII

THE LAKES AND SWAMPS OF DUNE LAND

TO THE south of the windward dunes many lakes are easily found. Some are small, with high, forested hills for boundaries. They all are irregular in shape. They are the spots that the traveling dunes have neglected, and are at or below the level of Lake Michigan. Long Lake, which is the longest of the inland lakes, varies in depth with the season and weather. Sometimes if the rainfall is above normal it is a lake throughout the year, but if there happens to come a dry season it is a bog or marsh. It is changing slowly by a filling-in process at the water's edge, and the grasses, shrubs, and trees from the higher land march steadily into the water. The water lilies which float the beautiful blossoms and lily pads, have to yield possession in favor of more stable vegetation.

The fires, that almost always sweep these inland lakes and swamps in the autumn, aid in making the new land. The ashes and residue from the flames are no small factor in filling in; the fire destruction being small compared with the resulting vigor to roots and the growth of the next season. Forest fires have probably done more good than harm in the dunes, considered as an aid to vegetation and land building.

A silt is being constantly blown over the land; some of it lodges, and a humus of decayed vegetation forms in time. A few hundred feet south of the electric line station at Miller can be traced the boundaries of an inland lake. The drainage of this land was occasioned by the laying of sewers in the village. The humus that had formed as a sediment when the lake existed was sufficient to beguile a farmer to plow and cultivate the land for a crop of corn. The experiment was only partly successful, the anaemic-looking stalks which came later told the story of a poor soil.

Walking through the trails one often sees broken branches of flowering shrubs and trees tossed aside, or clusters of wild flowers that have been torn up root and branch and dropped with the careless abandon of a burglar who has no use for the stolen goods. Frankly, of what use is a whole armful of flowers? Why the sacrifice to transient greed? A fellow dune lover once told me that while he admired the flowers with a passion for possession, he never picked a flower on any of his many trips. I overheard a globe-trotter, one who had gone around the world, say, "What one possesses is what one cares least for." It is true; an armful of flowers soon becomes a burden, and there is no sense in the wanton destruction. Gather a bloom if you will,

A first effort to cultivate the Dune Country. This corn field was a marsh some years ago, but recently village improvements at Miller, which is near-by, caused the water to disappear. The corn, however, is of a poor grade, and evidently not sufficiently nourished by the soil.

but cut the stalks and care for the blossoms afterwards and adorn your home with them. The cry of the dunes and marsh campaigners should be, "Spare the flowers, spare the vines, spare the ruthless destruction of tree or plant; do not tear down in a thoughtless moment the growth of several years. It is better to spare than destroy, and you are happier when you do."

In the early springtime in Japan many of the people make pilgrimages to the fields and woods where the blossoms are to be found. It is the almost universal custom to select some flower or shrub or tree, and to plead for its protection and to call attention to some characteristic beauty. To this end when a plant has been selected, a little piece of rice paper, three inches long and one inch wide, bearing some description, is fastened to a stick by pushing it through the long way of the paper. It is then placed against the tree or shrub or bloom to call particular attention to it. The inscription has individual ways of praying for its protection, and usually it reads, "I found this flower first and admired it; will you also enjoy it and protect it?" Ten thousand printed appeals on cloth and used in a similar way would protect ten thousand dune plants that are now ruthlessly destroyed.

On the edge of the swamps certain flowers may be found which rise to life year after year, and are to be seen nowhere else in the dunes. There are also certain

A swamp in autumn. The cat-tails are ripe, the water has gone, the willows and sedges will march down from the side hills and in a very few seasons the marsh will be overgrown with shrubs and trees.

zones for grasses, some of which must grow in the waters and others on the very margin of the land. It is hardly necessary to tell here of the various rushes or of the lily pads and water lilies. They are all there in the swamps, and the wild rice too is growing and so also the eel grass. How often I used to wonder why its name. The eel grass, which I came to know later was not a grass but a flowering plant—

> The flowers of which are small and are hidden in the sheath-like base of the leaves, which in narrow ribbons, a quarter of an inch wide, wave in long, thickly matted streamers in the channel beds, and shelter in their shady forest grass snails and worms and crabs and eels.[1]

The best time to visit the swamps, which at times seem to have the characteristics of marshes, is the midsummer season. Life then is at its crest. The promise of the spring is then fully completed and one feels that Nature is rejoicing over the fact that the summer for which so much preparation has been made is now here.

My first visit to the dune swamps was in the fall of the year. I got off the train at Long Lake and was told by a native, "Go down the railroad a piece, crawl through the wire fence, and take the path and don't get lost." My walk was only seven miles, but it was one of those yesterdays at the dunes which are fixed in my memory. The west end of the lake had dried so that

---

[1] C. W. Townsend, *Sand Dunes and Salt Marshes*, p. 191. Dana Estes & Co., Boston (The Page Co.).

A peach orchard is being submerged by a wandering dune. The tree in the foreground is heavily laden with peaches, although its trunk and many of its branches are under the sand. The photograph was taken a little distance south of the Tremont station on the electric line.

the path was all that was necessary to carry me dry shod into the dunes. My first find was a small, dried-up inland lake, filled with cat-tails and brown rushes, well ripe. My coming frightened a whole army of soldier blackbirds into flight. The camera snapped this pretty view of reeds and rushes, and the cat-tails had grown so rank that they had burst open, and as they pushed their heads high in the air each one seemed to be wearing a helmet and a plume.

The admonition "don't get lost" was kind, but it was unheeded, what did it matter? The spirit of the place was the same as the idea of the Indian who says, "No, I not lost, wigwam lost."

The dune swamp lands are beautiful in the fall. I think they do not lose by comparison with the salt marshes, so beautifully pictured by Dr. C. W. Townsend.

> In the fall of the year the marshes take on the yellowish-brown color, varying in different lights from silvery yellow to russet brown, threaded in places with basis of light pea-green, in places with dingy red, while the sapphire, hitherto invisible in its common green, blazes out in broad crimson patches. Still later in the year, before the ice and snow covers it all, the marshes wear a uniform russet-brown livery, a restful neutral-brown shade, a shade that hair and skin and clothing alike of all marsh dwellers assume in time.[1]

The universal law of change may be seen going on in the dune swamps; for example, along the Mineral

---

[1] *Sand Dunes and Salt Marshes*, p. 201.

Springs road directly north from the electric line, all the lowlands are emerging from undrained lakes into a vast swamp, and the swamp itself is fast filling, so that farmers are cultivating the newly made land. Beneath the surface, if one cares to know it, there are

The land vegetation is advancing into the marsh.

vast peat beds. The peat can be observed in the soils thrown up in building the drain ditches.

Farther along on the same road a most beautiful tamarack swamp commands attention and admiration. A botanist friend once said to me, "These ten acres of tamarack swamp are unequaled for beauty by anything anywhere in the dunes."

Those familiar with the development of marshes

and swamps and bogs, tell us that there are zones of different flora, that approach from the land down to and into the water. It is an orderly procession of plant life, and there are orderly changes in the margins and centers of the families of aquatic plants also. The first

On the edge of the peat bog at Dune Park.

plants in a pond are able to exist with little or no change in the water, except from rains and evaporation. The pond lilies and grasses may die, but their remains accumulate with almost no decay, and so the pond becomes shallower and finally is filled to the water's edge with undecayed vegetation. The top has plenty of water, but the deeper parts are undergoing compression. In time, owing to the advance from the land side of plants, which belong to the margins only, such as the bulrush and the sedges, the one-time lake changes to a swamp, and where the vegetation is undergoing partial decomposition, the swamp has all the characteristics of a bog.

Peats have a consistency varying from a slime to a turf, so in different seasons the various swamps appear in different forms owing to the presence or absence of moisture. Shrubs such as the swamp blueberry, cranberry, alders, willows, herbs, sedges, and mosses follow the aquatic plants as the water of marsh or bog disappears.

DUNE LAND AND HOW TO GET THERE

The Grand Calumet River at Miller. The ice is melting and the first blush of spring is showing in the budding cottonwood seedlings.

CHAPTER VIII

DUNE LAND AND HOW TO GET THERE

THE sand dunes of Indiana are located in La Porte, Porter, and Lake counties. The three counties are in the northwestern limit of the state, and for forty-three linear miles the southern beach line of Lake Michigan is the northern boundary of the state, and it is also the boundary of Dune Land.

The southern boundary of the state of Michigan is five and a half miles northeast from the signal station in Michigan City. The sand dunes skirt the southern boundary of Lake Michigan along this beach line for this same distance. Westward from the signal station in Michigan City, La Porte County, to the private pier light of the Indiana Steel Company in Lake County, is a distance of twenty-three and one-third miles. The north boundary of these counties is the south end of Lake Michigan. For a distance of twenty-eight miles the dunes parallel the sweep of Lake Michigan, extending from the west at Gary, in a northeasterly curved line, to the southwestern end line of Michigan. The active sand dunes are next to the water's edge the entire distance. The fixed sand dunes extend with varying distances directly south, ranging from three to fifteen miles. The Dune Lands comprise this large territory, and afford the greatest variety of typical sand dune landscapes. The place to see the active dunes is along the shore line of Lake Michigan.

Popular interest in the Dune Lands has been awakened recently, and an accompanying demand has come for information as to where the intending visitor may find the places of greatest charm and interest and how they may be reached.

HOW TO GO

The interesting places to visit in the Dune Country are centered around Miller, Dune Park, Mineral Springs, Chesterton, Tremont, Tamarack and Michigan City. Four trunk railroads run daily accommodation trains eastward to stations in dune land from Chicago: The Lake Shore & Michigan Southern R. R., the Michigan Central R. R., the Baltimore & Ohio R. R., and the Pennsylvania R. R.

The Chicago, Lake Shore and South Bend Railway Company, called the South Shore Lines, and known throughout the Dune Country as the electric line, operates twenty-two trains daily, between Chicago and South Bend, Indiana. All South Shore trains arrive at and depart from Randolph Street Station (Illinois Central R. R.), stopping for passengers at Van Buren Street, Central Station (Twelfth Street), Forty-

Carr's Beach (Main street).

The picnic trail at Carr's Beach.

third Street, Hyde Park (Fifty-third Street), Woodlawn (Sixty-third Street), Grand Crossing (Seventy-fifth Street), Kensington and Gary. The distance from Chicago to Miller is thirty-five miles; Dune Park, forty-one miles; Mineral Springs, forty-five miles; Portchester, forty-six miles; Tremont, forty-seven miles; Tamarack, fifty-one miles; Michigan City, fifty-six miles. Several intermediate stops are made at distances from one to three miles apart. The running time to Miller is one hour and thirty-one minutes, and to Michigan City two hours and fifteen minutes.

The automobile route to the dunes is over excellent roads. From Chicago to Miller, starting at the corner of Michigan and Jackson boulevards, the distance is thirty-six miles. Drive through South Chicago, Ninety-second Street and Buffalo Avenue east, turning south to the dyke across Lake George to Hammond, to Calumet, to Gary, 31.9 to Miller, north to Grand Calumet River 36.1. For places east of Miller take the St. Joseph road at Miller. At the first railroad crossing, 34.9 east to East Gary 36.6 to Waverly Beach; big sign at sharp turn of the road, 48.2; north two miles to Lake Michigan. Automobile routes from Chicago Heights, Dyer, Crown Point, Valparaiso, Indianapolis, South Bend, and LaPorte, from places east, south, or west, are through either Michigan City or Gary, Chesterton or Tremont.

Shore view in winter.

Another view of the shore when winter holds sway.

The old stage route, for many years the only road through this country, has been changed. In many places it may be followed by those who drive light motor cars or horse and wagon, and who prefer to journey leisurely over this old interesting dirt road.

WHAT TO SEE

Miller is an old town, and is easily accessible. The Grand Calumet River is the dividing line between the scrub oak forested plain at Miller and the sand dunes. Street improvements have made the approach to the dunes very easy. The direction from the various depots to the Grand Calumet River is due north, and the distance is about one and one-half miles. A new bridge spans the river and a board walk leads across the sand to Lake Michigan.

The chaotic appearance of the dunes from the approach to the bridge is in reality an orderly arrangement of the sands, that for years have been fashioned by prevailing winds. A most entertaining view of complex dune formation, standing and fallen trees, shrubs, vines and grasses, is from the south end of the bridge; while to the east the river widens into an inland lake, which heads almost a mile away. It is a picture spot, if one delights in banks, water and marsh views. To the west about a mile the river narrows to a width of ten or twenty feet; it has a very lazy

Waverley Beach and Dune Creek.

The home of "Diana of the Dunes."

current, and again it widens into a larger lake, which the steam dredge in 1906 helped to fashion into the present shore lines. At one time, centuries ago, this river found an outlet into Lake Michigan at this point, but drifting sands filled it and refashioned the river banks as we now see them. The great bank of white, clean sand, high above the waters' edge, with a steep convex surface, is another accumulation of sand of many years; it is constantly being blown over, and is slipping into the river; and from measurements that have been taken during the past fifteen years, it is estimated that it will take about a hundred or possibly a hundred and fifty years for the sand to fill the river.

The path westward from the bridge, on the south bank of this river, for a distance of three miles or more, is interesting to follow. Across the river may be seen long sweeps of variously fashioned dunes of queer conformation. Some are forested with both living and dead trees. Unique contours, contrasting colorings, curving ridges, and great banks of unbroken pure white sands, flowing downward by gravity, unless they are hindered by some obstructing tree trunk or tuft of grass, are here in profusion.

Carr's Beach is at the end of the board walk leading north from the bridge at Miller. It has a frontage of two miles on Lake Michigan. It has been known as Carr's for forty years. It has been and is now a favorite

When the ice is going out.

place for those who want the camp life. There are fifty houses and as many tents that are owned by, or rented to, the resident summer colony. The homes, such as they are, are placed, squatter fashion, without regard to any building lines. The beach itself is the

Trail from Miller to Carr's Beach.

only street. At Carr's a thousand visitors have enjoyed the bathing, not only one day, but on several occasions, and ten thousand people have taken advantage of a national holiday to visit this beach.

Dune Park is by far the best place to study the living dunes in all phases. One must walk to see it, and take all day if he can. The steam dredges have made sad ravages here, but it is still a wonder-land. Let your

The overwhelming advancing sand dune. Upon the white surface of the slope may be seen a new streak of sand of a slightly darker shade deposited during a severe rain storm. The vegetation in the foreground will soon be destroyed by the resistless mass of sand.

objective point be about a quarter of a mile to the west of the dredges at the water's edge. Notice here the three beaches. That nearest the water is the summer beach which, because of its being exposed to the action of the waves and the heat of the sun, is devoid of all life, with the exception of insects. The next is the middle beach, containing some debris and here and there a hardy annual plant. Above this is the upper beach, which is beyond wave action, and so it contains annuals, biennials and perennials.

The winters' gales have torn an opening through the high dunes here, leaving an open draw, almost level with the winter beach line. To the east and south the lee sands are spread in a broad, level plain, with desert characteristics and wind ripples of the largest type. To the east is the remnant of a dune said to have been one hundred and sixty-four feet in height, but now despoiled by the steam shovels. Out of this sand tomb many pine trees stand erect, though dead, having been buried for many years until exhumed by the dredges.

To the west there are many turreted or cone-shaped dunes. One of these is one hundred and forty-two feet high, and its steep windward side has a ragged planting of pine trees, irregularly grouped, all of which are weather-worn and many are broken. Here, at the roots of these pines, we find the dune form of heaths, consisting of cone-bearing dwarf shrubs, some growing

The coarse sand grass in the foreground is called *calamagrostis longifolia* by botanists, and it grows in exposed places where nothing else can grow. It is a hardy perennial resisting successfully summer heat and winter cold. In all, there are about thirty species of dune grasses.

erect, others sprawling on the sandy surface. These serve the double purpose of holding the moisture at the roots of the pines, and as dune binders. Here are dwarfed young junipers. The arbor-vitae and cedars are fairly numerous, and bend their boughs in a direction opposite the prevailing winds.

At some favorable place, climb one of these lakeward slopes to the crest and find how abruptly it ends in a sharp ridge hardly wide enough for you to walk upon. Before you, in the lee of this knife-blade crown, you will see a wide sweep of forest trees. The merciless sands are overwhelming the giants of this forest. The scattered pines, junipers, cedars, and oaks are sharing in the same destruction; their tops are scarcely showing above the sands. Continuing southward, cross the desert, and perchance you may find a small stone, triangular in shape, with smooth sides and edges, well rounded by sand friction. It is peculiar to a desert, and is called a dreikanter.

Next you will cross a plain, covered with rapidly growing alders, willow, cherry, and, in some spots, giant ragweeds. A few hundred feet west of the electric line station at Dune Park there is a chance to study all the sides of a small pond, with the shrub zone of plants in perfect type, such as swamp blueberry, cranberry, alders, dwarf birch, poison sumac, sedges, bullrushes, spirea and leatherleaf. To the east of this

The lone pine tree to the left is dead. A companion tree has been blown down in the center by the undercutting of winds. To the right the various grasses have not only held the dune but as the roots grew more tangled the gales were unable to make any impression upon it. A grape vine clinging to a tree with grasses at the roots was the probable nucleus.

spot, a few hundred feet, is a quaking bog, the water keeping the vegetation in a spongy and unstable state because of the lack of ordinary soil. It will be a peat bog when it is better drained. This very bog has a wealth of flora. A half-mile to the south from the Dune Park station, both east and west, are the marginal tree growths on the banks of the Little Calumet River. This flood-plain portion of the Dune Land is of great interest to botanists and ecologists.

Mineral Springs comes next, and the road is well paved up to the railroad crossing, and it then is a dirt road into the dunes. The dirt road divides into three paths. The one directly north will take you to the lake. To the east the path winds through to the lake a mile or so over the crest of a steep and high-sloping dune. The path to the west leads into a wonderful tamarack swamp. It is here the flora are in a perfect riot, according to the season. The brakes and cinnamon ferns are hip-high; pitcher plants, several varieties of orchids, many violets, purple asters, goldenrod, daisies, bottle and fringed gentians once grew in plenty near-by, but, alas, careless flower pickers were not content to pluck the blossoms, but must needs uproot the plants.

Port Chester road is the next crossing. The name, according to local tradition, was given to describe the way to Porter and Chesterton. It became "the Porter and Chester road," and, in daily use, "Portchester."

This picture shows all that is left of a once fairly large dune. The sand is being held by the roots of the dead trees, with the help of other vegetation.

The road to the beach here is due north from the electric line, and in a sharp curve joins the Tremont road. Together they form the direct road to the Waverly Beach. There is also a dirt road through the woods and over a most interesting complex dune of lee sands,

Tower on Mount Tom built by the Boy Scouts of Koutz, Ind.

through a forest and over a large dune of the horseshoe variety. There are types of all the dunes in this walk to the lake, both freaky and regular. When the visitor's time is limited, this is a good place to go to see the dunes.

Chesterton is a place that should be visited. I quote what Professor Cowles says about it:

All things considered, Chesterton is the most interesting place in the vicinity of Chicago, since it shows types of nearly all plant societies discussed in this paper: All phases of the river series from the ravine to the flood plain: all stages from pond to prairie: all types of dune activity, past and present, and morainic forests: spring brooks are as shown at this point, also the beech-maple forests.[1]

Tremont is next. The three conspicuous mountains of sand can be seen to the north. The one farthest west,

The approach to the beach-house of the Prairie Club.

one hundred and ninety-two feet high, is Mt. Tom. Tradition has it that a sailor whose name was Tom was buried at the top. The middle one is named Mt. Holden, in honor of the first president of the Prairie Club, and the one to the east is Mt. Green, in memory of a pioneer who kept the hotel near this station. The

[1] Henry C. Cowles, Professor of Plant Ecology, The University of Chicago. *The Plant Societies of Chicago and Vicinity*, p. 75. The University of Chicago Press.

Prairie Club has chosen for its beach-house a location near the foot of Mt. Tom. The marine surveying station on Mt. Tom is used as an observatory lookout by the public. The first tower was destroyed, and the present one was erected by the Boy Scouts of Koutz, Indiana. The views are magnificent. The panorama view of water and sand, forest and marsh, farm house and meadow, is one of charm.

From this vantage place the views of the shore side of the dunes are very attractive. Here one can watch the play of colors upon the surface of the great lake, which, in turn, awaken fancies or suggest wonderful pictures. Then there are the restless waves, breaking upon the wet sands with either a murmur or a roar. The walks and old wagon trails through the dense forest on the south side of these sharp crested dunes are a delight to those who know them well. To visitors they occasion great surprise and pleasure.

Waverly Beach is nine miles west of Michigan City by survey, but eleven miles by tradition. It is near the foot of Mt. Tom. It is a half-way place between Miller and Michigan City. The road to the water's edge is now graded, and is expected to be stone-dressed this year. Like Miller, it has its summer colony of transient visitors. Two thousand people a day have enjoyed this beach on various occasions, and during the pageant, June 3, 1917, from ten to twenty thousand began their

The lee side of an active dune showing three tree tops in the distance rising above the sand. Everything else in the way of vegetation has been covered by the dune. The trees in the foreground will most likely meet the same fate. Photograph taken near Michigan City.

acquaintance here with Dune Land. The fishermen tell of a row of sunken spiles and of a wooden pier that extended out into the lake six hundred feet, to take care of the shipping of lumber cut in the sawmill at this point. The little stream spanned by the new iron bridge was called in turn Fort Creek, City West Creek, and Coffee Creek, because of its iron-stained waters. More recently it has been named Dune Creek. A little town of forty houses and the big mill and three hotels constituted City West. Little Fort, a trading and a military post, was located on the high bank of Dune Creek. Mr. George A. Brennan, the historian, has a very interesting lecture, "The Wonders of the Dunes," which he frequently gives. It is historical, and centers around the description of the trails and old wagon roads near this spot. It is to be published in book form.

Tamarack is the station four miles further east. The train will leave you in a wilderness; the way out is by walking west half a mile, then turn north over a newly-made road covered with cinders. The route into the high dunes is by the old wagon trail. This part of the Dune Country is regarded by some as the best of all. Two miles to the west is doubtless the finest old forest of standing and fallen dead pine trees. There are parts of a very large blowout that are being refilled, so the phenomenon of a hundred years past, of dune activity and rest, are to be seen in a single hour's

The lace-like branches and spring buds of a cottonwood showing on a background of white sand. The tree shows a hard existence in the sand. The dune is a part of a blowout to the west of the board walk at Miller.

tramp. Once at Tamarack, it is a choice of going west to Waverly Beach or east to the Hoosier Beach near Michigan City; the distance is about five miles either way.

From Michigan City dune visitors may get off the electric line at Johnson's and go directly north into the traveling dunes which, during this very summer, have

The last remnant of what was once a beautiful group of trees.

been changing very rapidly. No more conspicuous sand piles can be found in the whole coast line than in this vicinity. Two high crests are of special interest —one rising to a height of one hundred and forty feet, and the other, due west a half-mile, to one hundred and twenty feet. The angle of the windward slope of the dunes, the beach, the water, the vegetation, and the drifting sand, moving often because feeling a zephyr's breath, make this place attractive.

Sheridan Beach is very popular with dune lovers. It is reached through the City Park at Michigan City. The drive to the east for three miles is over brick-paved roads. The dunes here, according to the United States Marine Survey charts, rise to a height of about seventy-six feet. The beach has on it many beautiful and expensive houses, with all modern improvements. This beach has taken on all of the characteristics of a summer resort, including week-end visits during the winter months. It is a restful, healthful and altogether a very charming spot in the heart of the dunes. The sand hills here are gray and can be seen for long distances out on the lake. While the dunes have been recently discovered by the public, thousands of people living near their boundaries have known them well for many years. Generations have enjoyed them, and their children's children love them.

Spotted something that isn't quite right?

Let us know and we'll fix it for future reprints.
And as a thank you for your diligence, we'll
send you a free Sothis book of your choice.

publisher@sothis.press

This page intentionally left blank

Except of course for that line above. And this one.
And whilst we have your attention, perhaps you'd
like to follow us on Instagram @sothis.press. Go on.

Made in the USA
Monee, IL
05 November 2026

67807985R00097